A Call for Change:

Recommendations for the Mathematical Preparation of Teachers of Mathematics

Committee on the Mathematical Education of Teachers

The Mathematical Association of America

MAA Notes and Reports Series

The MAA Notes and Reports Series, started in 1982, addresses a broad range of topics and themes of interest to all who are involved with undergraduate mathematics. The volumes in this series are readable, informative, and useful, and help the mathematical community keep up with developments of importance to mathematics.

MAA Notes

9. Computers and Mathematics: The Use of Computers in Undergraduate Instruction, *Committee on Computers in Mathematics Education, D. A. Smith, G. J. Porter, L. C. Leinbach, and R. H. Wenger,* Editors.
11. Keys to Improved Instruction by Teaching Assistants and Part-Time Instructors, *Committee on Teaching Assistants and Part-Time Instructors, Bettye Anne Case,* Editor.
13. Reshaping College Mathematics, *Committee on the Undergraduate Program in Mathematics, Lynn A. Steen,* Editor.
14. Mathematical Writing, by *Donald E. Knuth, Tracy Larrabee, and Paul M. Roberts.*
15. Discrete Mathematics in the First Two Years, *Anthony Ralston,* Editor.
16. Using Writing to Teach Mathematics, *Andrew Sterrett,* Editor.
17. Priming the Calculus Pump: Innovations and Resources, *Committee on Calculus Reform and the First Two Years,* a subcomittee of the Committee on the Undergraduate Program in Mathematics, *Thomas W. Tucker,* Editor.
18. Models for Undergraduate Research in Mathematics, *Lester Senechal,* Editor.
19. Visualization in Teaching and Learning Mathematics, *Committee on Computers in Mathematics Education, Steve Cunningham and Walter S. Zimmermann,* Editors.
20. The Laboratory Approach to Teaching Calculus, *L. Carl Leinbach et al.,* Editors.
21. Perspectives on Contemporary Statistics, *David C. Hoaglin and David S. Moore,* Editors.
22. Heeding the Call for Change: Suggestions for Curricular Action, *Lynn A. Steen,* Editor.
23. Statistical Abstract of Undergraduate Programs in the Mathematical Sciences and Computer Science in the United States: 1990–91 CBMS Survey, *Donald J. Albers, Don O. Loftsgaarden, Donald C. Rung, and Ann E. Watkins.*
24. Symbolic Computation in Undergraduate Mathematics Education, *Zaven A. Karian,* Editor.
25. The Concept of Function: Aspects of Epistemology and Pedagogy, *Guershon Harel and Ed Dubinsky,* Editors.
26. Statistics for the Twenty-First Century, *Florence and Sheldon Gordon,* Editors.
27. Resources for Calculus Collection, Volume 1: Learning by Discovery: A Lab Manual for Calculus, *Anita E. Solow,* Editor.
28. Resources for Calculus Collection, Volume 2: Calculus Problems for a New Century, *Robert Fraga,* Editor.
29. Resources for Calculus Collection, Volume 3: Applications of Calculus, *Philip Straffin,* Editor.
30. Resources for Calculus Collection, Volume 4: Problems for Student Investigation, *Michael B. Jackson and John R. Ramsay,* Editors.
31. Resources for Calculus Collection, Volume 5: Readings for Calculus, *Underwood Dudley,* Editor.
32. Essays in Humanistic Mathematics, *Alvin White,* Editor.
33. Research Issues in Undergraduate Mathematics Learning: Preliminary Analyses and Results, *James J. Kaput and Ed Dubinsky,* Editors.
34. In Eves' Circles, *Joby Milo Anthony,* Editor.
35. You're the Professor, What Next? Ideas and Resources for Preparing College Teachers, *The Committee on Preparation for College Teaching, Bettye Anne Case,* Editor.
36. Preparing for a New Calculus: Conference Proceedings, *Anita E. Solow,* Editor.
37. A Practical Guide to Cooperative Learning in Collegiate Mathematics, *Nancy L. Hagelgans, Barbara E. Reynolds, SDS, Keith Schwingendorf, Draga Vidakovic, Ed Dubinsky, Mazen Shahin, G. Joseph Wimbish, Jr.*
38. Models That Work: Case Studies in Effective Undergraduate Mathematics Programs, *Alan C. Tucker,* Editor.

39. Calculus: The Dynamics of Change, *CUPM Subcommittee on Calculus Reform and the First Two Years, A. Wayne Roberts,* Editor.
40. Vita Mathematica: Historical Research and Integration with Teaching, *Ronald Calinger,* Editor.

MAA Reports

1. A Curriculum in Flux: Mathematics at Two-Year Colleges, *Subcommittee on Mathematics Curriculum at Two-Year Colleges,* a joint committee of the MAA and the American Mathematical Association of Two-Year Colleges, *Ronald M. Davis,* Editor.
2. A Source Book for College Mathematics Teaching, *Committee on the Teaching of Undergraduate Mathematics, Alan H. Schoenfeld,* Editor.
3. A Call for Change: Recommendations for the Mathematical Preparation of Teachers of Mathematics, *Committee on the Mathematical Education of Teachers, James R. C. Leitzel,* Editor.
4. Library Recommendations for Undergraduate Mathematics, *CUPM ad hoc Subcommittee, Lynn A. Steen,* Editor.
5. Two-Year College Mathematics Library Recommendations, *CUPM ad hoc Subcommittee, Lynn A. Steen,* Editor.
6. Assessing Calculus Reform Efforts: A Report to the Community, *James R. C. Leitzel and Alan Tucker,* Editors.
7. Matching High School Preparation to College Needs: Prognostic and Diagnostic Testing, *John G. Harvey,* Editor and *Gail Burrill, John Dossey, John W. Kenelly, and Bert K. Waits,* Associate Editors.

Credit: Graph on page twenty-five reprinted with permission of the Middle School Math Project ©1990, The Math Learning Center.

Cover Image: Heleman Ferguson's *Double Torus Stonehenge: continuous linking and unlinking, $L \neq L$,* in twenty-eight pieces—one bronze and twenty-seven hard wax. "Some of the most important and pervasive ideas of twentieth century mathematics are the notions of continuous deformation, isotopy, and homotopy. This dance of the double tori celebrates that idea by taking a pseudolink, which most right-thinking persons would call a link and deforming it continuously without cutting or tearing into what every other right-thinking person would declare an unlinked state. Small and agreeable continuous perturbations relate each double torus to its neighbor; the total sum of these to the opposite number on the henge is large and surprising." Photograph courtesy of Noelle Ferguson.

These volumes can be ordered from:
MAA Service Center
P.O. Box 90973
Washington, DC 20036
1-800-331-1MAA FAX: 1-301-206-9789

ISBN 0-88385-071-0
Library of Congress Catalog Number 90-63690
Printed in the United States of America
Current Printing
10 9 8 7 6 5 4

Committee on the Mathematical Education of Teachers

Henry L. Alder, University of California-Davis, Davis, CA
Wade Ellis, Jr., West Valley College, Saratoga, CA
Henry Gore, Morehouse College, Atlanta, GA
M. Kathleen Heid, Pennsylvania State University, University Park, PA
Patricia L. Jones, University of Southwestern Louisiana, Lafayette, LA
Julie A. Keener, Central Oregon Community College, Bend, OR
James R. C. Leitzel, The Ohio State University, Columbus, OH (Chair)
Jo Ann Lutz, North Carolina School of Science & Mathematics, Durham, NC
Joseph Malkevitch, York College, CUNY, NY
Philip Wagreich, University of Illinois at Chicago, IL
Harriet Walton, Morehouse College, Atlanta, GA
Julian Weissglass, University of California-Santa Barbara, Santa Barbara, CA
R. O. Wells, Jr., Rice University, TX
Sue Poole White, 905 Sixth Street, SW, Washington, DC

Members of the Writing Team

Wade Ellis, Jr. West Valley College, Saratoga, CA
Marjorie Enneking, Portland State University, Portland, OR
M. Kathleen Heid, Pennsylvania State University, University Park, PA
Patricia L. Jones, University of Southwestern Louisiana, Lafayette, LA
James R. C. Leitzel, The Ohio State University, Columbus, OH

Acknowledgements

The writing team wishes to acknowledge the helpful suggestions received from mathematicians, mathematics educators, and teachers of school mathematics as this document evolved. It is not possible to list the many individuals who took the requisite time and effort to respond to the various drafts, but we do, in this way, extend our sincere thanks and appreciation for their thoughtful responses. We also wish to express appreciation and thanks to the Exxon Education Foundation for their support in the writing and dissemination of the document.

Special thanks are given to Ms. G. Denise Witcher and the Ohio State Department of Mathematics duplicating room staff. Given the time frame that constrained the writing team, their prompt and careful attention to detail were invaluable assets to the project.

Table of Contents

Acknowledgements	(vii)
Table of Contents	(ix)
Preface	(xi)
Introduction	(xiii)
Section I: Standards Common to the Preparation of Mathematics Teachers at All Levels	1
Section II: Standards for the Elementary (K-4) Level	11
Section III: Standards for the Middle Grades (5-8) Level	17
Section IV: Standards for the Secondary (9-12) Level	27
Section V: Closing Remarks	39
References	43
Appendix A: College and University Responsibilities for Mathematics Teacher Education	45
Appendix B: Previous MAA Recommendations	47

Preface

As we enter the decade of the 90's, the mathematics and mathematics education communities recognize and are responding to the call for dramatic changes in school mathematics. Two important documents have recently been published which, together, present a new vision for school mathematics: ***Everybody Counts: A Report to the Nation on the Future of Mathematics Education*** (National Research Council, January, 1989) and ***Curriculum and Evaluation Standards for School Mathematics*** (National Council of Teachers of Mathematics, March, 1989).

These documents address, among other items, such critical issues as the nature of mathematics, the need for change in emphasis and content in the school curriculum, and the role of calculators and computers in learning and doing mathematics. Particularly, they stress the importance of how mathematics is taught. NCTM's ***Standards*** state:

> ***How*** mathematics is taught is just as important as ***what*** is taught. Students' ability to reason, solve problems, and use mathematics to communicate their ideas will develop only if they actively and frequently engage in these processes. Whether students come to view mathematics as an integrated whole instead of a fragmented collection of arbitrary topics and whether they ultimately come to value mathematics will depend largely on how the subject is taught.

There is an overwhelming consensus that students of the 1990's and beyond will develop "mathematical power" only if they are actively involved in **doing** mathematics at every grade level. "Mathematical power" denotes a person's abilities to explore, conjecture, and reason logically, as well as the ability to use a variety of mathematical methods effectively to solve problems.

Such substantive changes in school mathematics will require corresponding changes in the preparation of teachers. In order for teachers to implement the curriculum envisioned by the NCTM ***Standards***, they must have opportunities in their collegiate courses to do mathematics: explore, analyze, construct models, collect and represent data, present arguments, and solve problems. The content of collegiate level courses must reflect the changes in emphases and content of the emerging school curriculum and the rapidly broadening scope of mathematics itself.

In general, current requirements for certification of teachers of school mathematics, particularly at the elementary and middle school levels, and the learning experiences of prospective teachers within college mathematics classes fall far short of these goals. There is a need for change. This document calls for change in

how and **what** mathematics is taught to prospective teachers in collegiate courses and change in the amount of mathematics preparation now required for the certification of teachers of school mathematics at all levels.

To change the teaching and learning of mathematics in the nation's schools, the preparation of teachers must also include developing an understanding of students as learners of mathematics, obtaining appropriate background in mathematical pedagogy, and constructing suitable classroom environments to foster learning by all students. Recommendations on these aspects of the preparation of teachers of mathematics are given in ***Professional Standards for Teaching Mathematics*** (National Council of Teachers of Mathematics, March, 1991).

The Committee on the Mathematical Education of Teachers continues the tradition established by the Mathematical Association of America's Panel on Teacher Training in making recommendations for the mathematical preparation of teachers (See Appendix B). The 1983 document, ***Recommendations for the Mathematical Preparation of Teachers*** (MAA Notes Number 2), provided a foundation upon which this current document builds. The earlier publication still contains much useful material for those developing programs in teacher preparation.

Introduction

The recommendations for the preparation of teachers of mathematics come from a vision of "ideal" mathematics teachers in classrooms of the 1990's and beyond. Such persons must:

- communicate mathematical ideas with ease and clarity;
- organize and analyze information, solve problems readily, and construct logical arguments;
- possess knowledge and have an understanding of mathematics that is considerably deeper than that required for the school mathematics they will teach;
- enjoy mathematics and appreciate its power and beauty;
- understand how mathematics permeates our lives and how the various threads within mathematics are interwoven;
- naturally and routinely use technology in the learning, teaching, and doing of mathematics.

This document describes the collegiate **mathematical** experiences that a teacher needs in order to meet this vision. Teaching mathematics is a complex task that requires more than just a sound foundation in mathematics. Other dimensions needed in the preparation of teachers of mathematics are addressed in ***Professional Standards for Teaching Mathematics*** (National Council of Teachers of Mathematics, March, 1991). Programs preparing teachers of mathematics must take into account the full range of recommendations contained in each of these documents.

The three levels of instruction referred to throughout this document are similar to those used in the ***Recommendations on the Mathematical Preparation of Teachers*** (MAA Notes Number 2, 1983) and reflect the descriptions given in recent NCTM publications:

Elementary (K—4) Teachers of early childhood and primary school mathematics

Middle Grades (5—8) Teachers of middle/junior high school mathematics; elementary school mathematics specialists, and coordinators of elementary school mathematics programs

Secondary (9—12) Teachers of high school mathematics

These levels of instruction are used as guides to frame the standards that follow. Local situations and custom will require modification and interpretation of these levels if the intent is to use this document to describe specific certification guidelines.

The first section contains six Standards which should be common to the experience of mathematics teachers at every grade level. The subsequent sections address standards for the mathematical preparation of teachers at specific grade levels. The Standards given in each section describe broad knowledge and understanding of mathematics needed by mathematics teachers.

It is not intended, nor should the document be read to imply, that a given Standard describes the content of a single college-level mathematics course. Realization of a given Standard will only be achieved through experiences provided in several courses in the collegiate curriculum. The changing nature of mathematics and the new opportunities to rethink content and instruction opened by developments in technology provide opportunity for dramatic changes in the collegiate curriculum.

NCTM's ***Curriculum and Evaluation Standards for School Mathematics*** envisions what the school mathematics curriculum should include in terms of content priority and emphasis and an elaboration of appropriate ways of evaluating student learning and school programs in mathematics. This document identifies the mathematical knowledge, experiences, and skills necessary to enable teachers to teach the program described in the ***Curriculum and Evaluation Standards for School Mathematics.*** For practicing teachers to become fully involved with professional development they will need support (both moral and financial) and time to participate. Local school districts become a part of the change process by actively encouraging and supporting teachers in their efforts to improve. State certification agencies become a part of the change process by reviewing their existing policies to assure that they reflect current statements on professional teaching standards and the depth of mathematics study presented in these recommendations. Mathematics is so important a discipline that its teaching, at every grade level, must be assigned to the best prepared individuals.

In this document we speak directly to changes recommended in the collegiate mathematics curriculum designed for the preparation of teachers. Since in most institutions prospective secondary teachers take the same mathematics courses as other mathematics students, this document addresses issues that affect the entire undergraduate major. In this time of exciting developments in the mathematics profession, we who teach mathematics in colleges and universities have an important role to play as agents of change. As we have been successful in meeting past challenges, so we will also work toward the new goals for curriculum revision. While standing independently as a statement to the mathematics community, this document also represents the contribution of the Mathematical Association of America to the cooperative efforts of major organizations currently involved in the reform of mathematics education.

Section I

Standards Common to the Preparation of Mathematics Teachers at all Levels.

Standards in this section encompass the preparation recommended for mathematics teachers so that they (i) view mathematics as a system of interrelated principles, (ii) are able to communicate mathematics accurately, both orally and in writing, (iii) understand the elements of mathematical modeling, (iv) understand and use calculators and computers appropriately in the teaching and learning of mathematics, and (v) appreciate the development of mathematics both historically and culturally. The recommendations in this section will be assumed to influence, and be a part of, all mathematics courses taken by prospective and practicing teachers of mathematics.

In developing programs for teachers of mathematics that incorporate these standards, we assume that appropriate attention, interpretation, and emphasis will be given to the level of school mathematics for which the preparation is intended.

The standards in this section are:

1. Learning Mathematical Ideas
2. Connecting Mathematical Ideas
3. Communicating Mathematical Ideas
4. Building Mathematical Models
5. Using Technology
6. Developing Perspectives

Learning Mathematical Ideas

Mathematics is learned by doing. In the process of exploring problems, designing and conducting experiments, making, testing, and refining conjectures, and proving theorems, learners begin to understand the value of inductive and deductive reasoning in mathematics. They develop skills in recognizing, extending, and verifying patterns, identifying assumptions, building systems, and using results. They learn not only to answer questions and solve someone else's problems, but to pose their own questions and problems. Through a variety of these learning experiences, mathematics teachers develop a sense of self-confidence and a willingness to explore and learn new mathematics on their own. To make this possible, collegiate mathematics classrooms must become a place where students actively do mathematics rather than simply learn about it.

Standard 1: Learning Mathematical Ideas

The mathematical preparation of teachers must enable them to:

- ***become independent learners, capable of doing and learning mathematics on their own;***
- ***develop their own processes, concepts, and techniques for solving problems;***
- ***exercise mathematical reasoning through recognizing patterns, making and refining conjectures and definitions, and constructing logical arguments, both formal and heuristic, to justify results.***

In collegiate classes, students must be active participants in the learning process rather than passive recipients of information. Classroom activities should include critical thinking and questioning, searching for patterns and contradictions, discussion, debate, and persuasive argument. Students should read and discuss text material and alternative sources, including expository and historical articles. They must experience the use of resource materials, including physical models and technology to enhance their understanding of ideas. They should explore new (to them) areas of mathematics independently and through cooperative teamwork. In such a classroom, the role of collegiate mathematics instructors will change from one of disseminating information to one of guiding the learning of mathematics. This change may not be easy, and for many of us it will take time and effort. ***A Source Book for College Mathematics Teaching*** (MAA Notes and Reports, 1990), prepared by the MAA's Committee on the Teaching of Undergraduate Mathematics, provides suggestions and resources to help expand a repertoire of instructional techniques.

Connecting Mathematical Ideas

The beauty and utility of mathematics often appears in the connections that exist among different branches of mathematics and between mathematics and other disciplines. Recognizing and exploring interrelationships among the various parts of mathematics enhances the learning of mathematics. These connections enrich the conceptual framework students use to understand the world and are important in motivating their continued study of mathematics. Teachers of school mathematics must also understand how mathematics permeates every facet of modern society. Its importance in nonscientific disciplines, and in the everyday life of all of us, is not generally understood; and, if this is to change, prospective and practicing mathematics teachers need many opportunities to formulate, pose, and solve problems arising from a wide variety of sources.

Standard 2: Connecting Mathematical Ideas

The mathematical preparation of teachers must provide experiences in which they:

- ***develop an understanding of the interrelationships within mathematics and an appreciation of its unity;***
- ***explore the connections that exist between mathematics and other disciplines;***
- ***apply mathematics learned in one context to the solution of problems in other contexts.***

In college courses we should look for and highlight the richness of mathematical situations in which a concept can be represented in several ways. We should pose problems that can be explored through a variety of approaches. We should emphasize concepts and processes developed in other courses that are helpful or illuminating. Students must regularly have opportunities to explore and exploit the connections among various representations of mathematical concepts and compare their usefulness in a particular problem.

Teachers need to recognize the relationship between what they teach and what is taught at other levels of school mathematics. They need, for example, to understand the close parallel among the development of integer arithmetic in the elementary grades, the algebra of polynomials in the middle and early high school curriculum, and the ideas of number systems explored later in high school. They should see that counting processes and the concepts of functions and relations permeate all aspects of mathematics. They should explore the relationships between geometry and algebra and the use of one to investigate the other.

Their experiences in mathematics classes must develop these relationships and show that the topics are best learned when not isolated from each other.

Striking applications of mathematics occur not only in the physical sciences, but also in areas as diverse as business, politics, and social sciences. Problems in traffic flow, product distribution, communications, epidemiology, and natural resource management provide examples of areas in which mathematics is heavily used. The formulation of a problem in mathematical terms frequently rests on the insights developed through the connections among various areas of mathematics and between mathematics and other disciplines. Mathematics teachers need to have an appreciation of that interplay through their own experience.

Communicating Mathematical Ideas

Mathematics teachers at all levels must be able to communicate mathematical ideas with ease and clarity. Teachers must discuss concepts, ask and answer questions, guide explorations, pose problems, present logical arguments, and critique the work of their students. Communicating mathematics is also a necessary part of their own learning experience. As they struggle to phrase a definition or justify a conjecture or describe a problem situation, students deepen their own understanding of the underlying mathematics.

Standard 3: Communicating Mathematical Ideas

The mathematical preparation of teachers must enable them to:

- ***develop skills in both written and oral communication of mathematical concepts and technical information;***
- ***learn to communicate effectively at various levels of formality and with people who have differing levels of mathematical insight;***
- ***understand and appreciate the power of mathematical language and symbolism in the development of mathematical concepts.***

Teachers will attain this standard as they are routinely expected to talk and write about mathematics in their college mathematics courses. They need opportunities to work in groups to construct models or solve problems, to draw figures to clarify problem situations, and to design graphs to represent sets of data. We should regularly ask them to respond to questions which promote discussion or require considerable explanation. (For example,

How do you know this solution is unique? Why are all measurements approximations? Why do you believe this conjecture is correct? To verify that p is a prime, why is it sufficient to consider divisibility of p only by primes less than or equal to $\sqrt{p}$?) We should require them to present written and oral arguments and explanations, justify their own reasoning, critique the reasoning of others, and answer discussion questions on tests. To facilitate the development of communication skills, we must pay attention not only to the mathematical correctness of our students' written and oral work, but also to the quality of such exposition, and provide, on a regular basis, the kind of feedback which will help students improve.

Building Mathematical Models

There are three ways in which models are used in attempts to solve problems originating in the world around us:
(i) information is derived from mathematical models that others have built and used (weather reports, governmental studies, stock market projections); (ii) models that others have derived are used to analyze real-world situations; and (iii) people derive their own mathematical models of these situations either from known quantitative relationships or from collected data.

Mathematics teachers should have experiences using models at all three of these levels. Formal exposure to the power, limitations, and consequences of particular mathematical models will sensitize teachers to the benefits and pitfalls of using mathematics in the real world. They must understand certain very fundamental principles underlying the concept of modeling and have experience constructing their own models at appropriate levels of difficulty. In particular, every teacher of mathematics should understand that models at best only approximate reality since they are constructed on the basis of assumptions (however well conceived); there are frequently limitations on the applicability of mathematical models; and models need to be checked against known and reasonable values in the context of the original problem.

Standard 4: Building Mathematical Models

The mathematical preparation of teachers must include experiences that enable, motivate, and encourage them to analyze real-world situations through the use of whatever mathematical ideas or quantitative strategies are available. In particular, they should be able to:

- *work with a given model;*
- *recognize constraints inherent in a given model;*
- *construct models to analyze real-world settings and use symbols and reasoning in analysis;*
- *convert among representations (graphical, numerical, symbolic, verbal) that reflect quantitative constraints in a given real-world setting.*

Course work should develop fundamental understanding of the modeling process rather than make teachers experts on particular types of models. We expect experiences with mathematical modeling techniques to be a part of most courses in a collegiate mathematics program. These experiences should allow mathematics teachers to feel comfortable in approaching a variety of problems unfamiliar to them. It is important that their experiences include using multiple representations, interpreting models, and checking the results predicted by the model against reality.

Mathematical modeling software should become an integral part of the education and classroom experience of mathematics teachers. Currently available software ranges from simulations of tossing dice to sophisticated simulations of the flow of blood, air turbulence over aircraft, and the flight characteristics of flocks of birds. Teachers should have experience with such modeling programs, both in the classroom and in laboratory settings.

Using Technology

Current, available technology has changed the practice of research mathematics and is profoundly changing the teaching and learning of mathematics at all levels. Given carefully designed instructions, computers are able to aid in visualizing abstract concepts and to create new environments which extend reality. We risk limiting students' mathematical power by divorcing mathematics from technology. Greater accessibility to mathematical strategies and representations provided by a technologically rich environment opens a new array of real-world problems to mathematical solution. Calculators and computers allow students to explore mathematical ideas from several different perspectives, resulting in a deeper mathematical understanding. Through the regular use of calculators and computers in collegiate mathematics courses, students learn more mathematics and can more rapidly apply that understanding in problem solving. The strategies and representations encouraged by use of technology are at present largely untapped both in undergraduate classes and school mathematics programs. Thus, there is reason to change what and how we teach prospective and practicing teachers. Students need experience in using appropriate technology effectively in solving problems so that they can learn and adapt strategies and representations that arise while using that technology. Because they can reflect on their own learning and understanding of mathematical ideas when using technology, teachers will be better prepared to lead their own students in effective mathematical learning using calculators and computers.

Standard 5: Using Technology

The mathematical preparation of teachers must include experiences in which they use calculators and computers:

- ***as tools to represent mathematical ideas and construct different representations of mathematical concepts;***
- ***to engender a broad array of mathematical modes of thinking through use of powerful computing tools (including function graphers, curve fitters, and symbolic manipulators);***
- ***to develop and use alternate strategies for solving problems.***

In the mathematics classroom, prospective teachers should use calculators and computers to pose problems, explore patterns, test conjectures, conduct simulations, and organize and represent data. Computer software and calculators that quickly and accurately perform numerical, graphical, and symbolic computations are useful as problem-solving tools and can enhance students' mathematical understanding. Teachers should have this experience in whole-group, individual, and small-group extended problem solving settings.

Although the real challenge is the fundamental reshaping of the content and processes of teaching college mathematics, we must recognize that in some settings this is necessarily a long-term goal. Yet, mathematics instructors can take the first steps now. While the emphasis may vary from time to time, there will be significant places in almost every college mathematics course where technology can find use. Inexpensive graphing calculators are available for immediate use in all undergraduate courses. The computer equipment (a computer, a monitor, and an overhead display device) needed for in-class demonstrations is also available. As colleges budget for, design, and build computer labs, we can assign appropriate computer-based activities to our students. Two- and three-dimensional function graphers, spreadsheets, data analysis software, simulation software, and integrated symbolic/graphical/numerical manipulators are excellent software packages on which to base exploratory problem-solving assignments. Computer-based materials for elementary geometry, probability, and statistics are currently on the market. Collections of such assignments designed for calculus courses are now becoming commercially available.

Developing Perspectives

Mathematics is a truly human endeavor. Its teaching should include a close look at the development of mathematical ideas and the women and men who have contributed to that development throughout history and who are playing important roles in mathematics today. These mathematical ideas should not be rooted solely in the past. We who are teaching collegiate mathematics should make our students aware of the striking new developments taking place today in mathematics and its applications. An understanding of mathematicians as people, with their wonderful diversity of personalities, idiosyncrasies, backgrounds, and interests is vital for those who will be teaching mathematics. We need to share our appreciation of the contributions of mathematics to society and the impact of mathematicians on society.

Historically, many results in theoretical mathematics developed from a need to solve problems in disciplines outside mathematics. Early techniques and empirical rules for measurement were followed much later by an axiomatic development of Euclidean geometry. Calculus was used to solve problems in the sciences for two hundred years prior to its theoretical development. The need for record keeping and the organization and interpretation of data in increasingly complex societies gave rise to the development of statistical theories. But mathematicians throughout history have also created or extended many areas of mathematics for the sheer enjoyment of exploring mathematical ideas, never dreaming that years later these mathematical ideas would become immensely useful to people in business and industry. (Graph theory, with its multiple areas of application to such topics as network development, transportation problems, and allocation of resources, has ties to recreational parlor games of the 19th century and to religious rites in cultures of Africa and the islands of the South Seas.) Throughout history the interplay of mathematics and other disciplines, as well as the interplay of different areas within mathematics, has been essential in the development of mathematics and continues to be so today.

Standard 6: Developing Perspectives

The mathematical preparation of teachers must include experiences in which they:

- *explore the dynamic nature of mathematics and its increasingly significant role in social, cultural, and economic development;*
- *develop an appreciation of the contributions made by various cultures to the growth and development of mathematical ideas;*
- *investigate the contributions made by individuals, both female and male, and from a variety of cultures, in the development of ancient, modern, and current mathematical topics;*
- *gain an understanding of the historical development of major school mathematics concepts.*

The breadth of this recommendation cannot be achieved through a single college course. Therefore, we are encouraging the view that historical and cultural aspects of mathematics be incorporated in all undergraduate mathematics courses. These experiences might include, but are not necessarily structured by, a chronological approach to the development of mathematical ideas. The emphasis should be on concepts and their interrelationships. It is also important that teachers of mathematics have opportunities to become acquainted with some current literature in mathematics and biographies of contemporary, 20th century mathematicians. Teachers should be aware of how the search for solutions to problems has created, and continues to create, new mathematical concepts and procedures.

The historical and cultural development of mathematical ideas related to the school mathematics curriculum is particularly relevant for teachers. They should understand the evolving use of mathematical notation and terminology and be able to trace the origins of the major topics in arithmetic, algebra, geometry, trigonometry, calculus, number theory, probability, statistics, discrete mathematics, and the development of calculators and computers and their impact on the teaching and learning of mathematics.

Section II

Standards for the Elementary Grades (K—4)

In grades K—4, teachers build on a child's primitive and intuitive ideas of number, shape, and size. Recognizing that the child's understanding is at a very concrete level, they will be teaching in an environment in which manipulative materials, calculators, and computers should be used on a regular basis as an integral part of the elementary mathematics program.

In order for teachers to engage students in such classroom activities, their college mathematical preparation must include a core of experiences that we describe within four broad standards:

1. Nature and Use of Number
2. Geometry and Measurement
3. Patterns and Functions
4. Collecting, Representing, and Interpreting Data

These Standards are not intended to define individual courses. They describe experiences to be included in the mathematical preparation for a teacher at the elementary level. To provide these experiences, a minimum of **nine** semester hours in content mathematics is needed, assuming as prerequisite three years of college preparatory high school mathematics or the equivalent.

The mathematical experiences recommended for teachers at the K—4 level require that mathematics departments offer courses specifically designed for this audience. In so far as their programs permit, teachers at the elementary level should be encouraged to select additional courses from those offered to meet the Standards described in Section III to place the mathematics they will be teaching in a broader context.

Nature and Use of Number

Elementary teachers must develop an understanding of the nature and use of numbers and an appreciation of their importance. They must recognize that almost every aspect of our lives is quantified in some way and that the ability to use numbers appropriately and effectively is essential for everyone. In brief, a well developed "number sense" is critical. This number sense consists of finding easy and alternative ways of doing computations, of knowing when and how to estimate, and when and how to get exact answers. It incorporates decisions of when to use, or not to use, calculators or computers and of knowing whether the number appearing on the display screen or in a print-out is reasonable. It permeates the understanding of patterns as well as issues in measurement.

Standard 1: Nature and Use of Number

The mathematical preparation for teachers of the elementary grades must provide experiences in which they:

- ***investigate the role of numbers as a logical, predictable system for expressing and relating quantities;***
- ***analyze and compare features and basic computational techniques in selected numeration systems in use today and in the past;***
- ***explore the operations, properties, and uses of whole numbers, fractions, and decimals;***
- ***use estimation and mental arithmetic, calculators, computers, paper-and-pencil algorithms, and manipulative materials, in solving a wide variety of problems.***

In mathematics courses for elementary teachers, we must provide opportunities for them to explore basic concepts and operations of whole numbers and fractions using a variety of physical models (e.g., multibase blocks, fraction bars, colored chips, and geoboards). These models can be used to investigate and solve many real-world problems that involve ratio, proportion, percent, simple probability, length, area, or volume. Estimation and mental arithmetic need to be emphasized throughout the mathematics courses for these teachers. Teachers should be led to develop an understanding of the fundamental differences between numbers which result from counting and numbers which result from measurements. Activities should include actually making measurements and discussing questions of precision, rounding, and significant digits.

A basic component in developing ideas about the nature of number is the study of properties of the whole numbers, including divisibility and factorization. Teachers should have the opportunity to use manipulatives, visual approaches, calculators, and computers to develop concepts of prime, factor, greatest common divisor, and least common multiple, and the significance of these concepts in the use of numbers. Elementary number theory is an excellent source of easily stated, challenging, and motivating problems which can often be explored with a calculator or computer. (Prospective and practicing teachers may, for example, investigate unit fractions and try to determine whether every unit fraction can be written as the sum of two different unit fractions. They may use a calculator to explore terminating versus repeating decimals and to determine which fractions have terminating decimal expansions. They may try to predict the maximum length of the repeating block of the decimal form of a given fraction.)

Geometry and Measurement

The study of geometry and measurement by prospective and practicing elementary teachers must be rooted in explorations and hands-on activities and motivated by problems whose solutions involve spatial sense. Teachers develop confidence and expertise in the use of units and measuring techniques by actually making and interpreting measurements. They develop a real understanding of geometric figures and relationships by constructing and exploring models.

Standard 2: Geometry and Measurement

The mathematical preparation for teachers of the elementary grades must provide experiences in which they:

- ***use a variety of tools, physical models, and appropriate technology to develop an understanding of geometric concepts and relationships and their use in describing the world in which we live;***
- ***make and interpret measurements of many kinds of two- and three-dimensional objects;***
- ***formulate and solve problems whose solutions require spatial sense.***

Elementary teachers should use physical and visual models whenever possible in their exploration of distance, congruence, similarity, area, volume, symmetries, and transformations. In their mathematical course experiences, teachers should routinely use pictures, tangrams, pattern blocks, geoboards, and appropriate computer software. For example, LOGO procedures could be used to explore ideas of symmetry and pattern. Teachers should use ruler, compass, and protractor to construct two- and three-dimensional figures and explore relationships among these figures. They should develop the meaning of area, volume, and capacity through the measuring of many nonstandard figures and objects, and distinguish between the meaning of these concepts and the formulas for specific items. They should explore properties of polyhedra, pyramids, cylinders, and cones. Questions of precision, accuracy, and appropriate units will arise as students explore similarity and measurement. These activities provide excellent opportunities for students to work in groups and discuss their discoveries, insights, and difficulties.

Mathematics teachers must learn to recognize patterns, make conjectures, and present heuristic arguments to support or explain their conjectures. Such results as the Pythagorean Theorem, Euler's Formula, the ability to tessellate with any quadrilateral, symmetry properties of certain figures, numerical relationships of similar figures, and formulas for measures of specific basic figures, provide examples of situations in which prospective elementary teachers can experience the process of discovery, generalization, justification of answers, and communication of ideas.

Patterns and Functions

Recognition and use of patterns and the concept of function are fundamental themes permeating the study of mathematics. Through the study of patterns which emerge from various problem situations, relations and functions evolve, as well as the language and symbolism used to describe them. Prospective and practicing elementary teachers must have opportunities in their collegiate mathematics courses to search for, describe, and generalize patterns in a broad range of problem situations. They need to recognize and describe relations and functions at an informal and intuitive level and through numerical, algebraic, and geometric representations.

Standard 3: Patterns and Functions

The mathematical preparation for teachers of the elementary grades must provide experiences in which they:

- ***recognize the study of patterns as an underlying, fundamental theme in mathematics;***
- ***create and use pictures, charts, and graphs to recognize and describe mathematical relationships;***
- ***discover and analyze functional relationships which arise from diverse problem situations;***
- ***develop the use of variables and other algebraic notation as an efficient and natural way to describe relationships.***

Patterns can be used to develop computational methods and number properties, to recognize and describe certain geometric properties, and to organize and analyze data. Visual descriptions of patterns, including pictures, charts, and graphs can be created both by hand and with the aid of calculators and computers. The properties of polygons, demographic studies, and results from food and drug testing are rich sources for problems whose solutions require the recognition and use of patterns. Through appropriate choice of problems, prospective and practicing elementary teachers can develop models, symbols, and language to describe the concepts of equality, inequality, symmetry, congruence, proximity, and rate of change. The use of variables will arise naturally as a way to describe efficiently these patterns, relations, and functions. Variables and functional relationships may be further explored with calculators and computer utilities.

The process of recognizing patterns, making, explaining, and justifying conjectures, and developing ways of discussing functional relationships provides students with the opportunity to enhance their use of mathematical language and symbolism and their reasoning and communication skills.

Collecting, Representing, and Interpreting Data

Graphs, charts, and other representations of statistical information abound in newspapers, magazines, textbooks, and the visual media. Elementary teachers must have experiences in data collection, representation, and analysis which will enable them to discuss and interpret these various forms of collected data with their students.

Standard 4: Collecting, Representing, and Interpreting Data

The mathematical preparation for teachers of the elementary grades must provide experiences in which they:

- ***collect and interpret data represented in different ways;***
- ***conduct sampling experiments to develop an appreciation for randomness;***
- ***explore empirical probability from data they have collected and relate it to theoretical probability based on a description of the underlying sample space;***
- ***explore and compare various methods for representing data, both by hand and by using calculators and computers.***

Elementary teachers should understand the concept of randomness and how bias can distort a sampling process. They need opportunities to design, conduct, analyze, and discuss sampling experiments and other data-collection techniques. They should explore various means of representing data from these experiments in order to decide what type of representation is most appropriate for a particular set of data. Experience and exploration with calculators and computers, particularly simulation strategies, as well as paper-and-pencil methods, should be integral to their study of these ideas. They should understand that mean, median, mode, and percentile ranking provide different ways to describe the central tendencies of collected data. Discussions of distributions of scores from SAT or other standardized tests can be related directly to these ideas.

Teachers should develop an understanding of probability in the context of equally likely sample spaces and in its application to such common things as weather reporting and the setting of insurance rates. All these activities provide excellent opportunities for learners to work in groups and discuss their discoveries, insights, and difficulties.

Section III

Standards for the Middle Grades (5—8)

The preparation of middle grades teachers is particularly crucial in the overall reform of school mathematics. Students at this age often make firm decisions about how and whether to continue their study of mathematics. Teachers at this level need to know how the mathematics they teach follows from elementary school mathematics and how it leads to the secondary curriculum. They need a breadth and depth of experiences which go considerably beyond the preparation of elementary teachers but which are quite different from that expected for teachers at the secondary level. The recommended core of such experiences is given in five broad Standards:

1. Number Concepts and Relationships
2. Geometry and Measurement
3. Algebra and Algebraic Structures
4. Probability and Statistics
5. Concepts of Calculus

A program which meets these Standards requires at least **15** semester-hours of content mathematics, assuming as prerequisite four years of college preparatory high school mathematics and should incorporate the experiences recommended in Section II for teachers at the K—4 level.

Even though many states have introduced certification or endorsements for the middle-grades level, specialized mathematics experiences for teachers at this level are not common. The breadth of mathematical experiences needed by teachers of the middle grades is enormous, but the depth of study appropriate for them is not necessarily the same as that expected for mathematics majors. For example, teachers of the middle grades should develop an intuitive understanding of limits, continuity, derivatives, and integrals. This is seldom acquired from the first course in calculus designed for those majoring in science or engineering. On the other hand, certain ideas, such as place-value concepts, intuitive geometry, and rational number arithmetic need special emphasis for teachers of the middle grades.

If the recommended Standards cannot be met satisfactorily within currently offered undergraduate courses, then special courses should be developed that provide the proper focus and breadth of experience for these teachers.

Number Concepts and Relationships

The ability to reject absurd answers, to estimate, to compare quantities, and to have numbers make sense in the myriad charts, graphs, measurements, and computer print-outs in our daily lives is a basic mathematical skill. This skill should be acquired in grades 5—8. Prospective and practicing teachers of the middle grades must have a basic understanding of numbers and their symbolic representations. It is particularly important for teachers of the middle grades to have a solid understanding of the conceptual basis for the numbers and operations with which their students will work.

Standard 1: Number Concepts and Relationships

The mathematical preparation of middle-grade mathematics teachers must include experiences in which they:

- ***develop a practical, concrete sense of number;***
- ***use physical materials and models to explore fundamental properties of number systems;***
- ***develop conjectures and intuitive proofs of number theoretic properties;***
- ***investigate number sequences, patterns, and functional relationships;***
- ***explore the meaning of infinity and its role in the study and historical development of topics such as geometry and calculus.***

Number theory experiences for middle-grade mathematics teachers should be characterized by concrete explorations of number systems and patterns. They should be designed to develop teachers' expertise in mental arithmetic, estimation, and determining the reasonableness of answers. These experiences, building on those already expected from the program for K—4 teachers, will provide the opportunity for more challenging explorations and extensions and further development of their "number sense." Physical models should be used to review the development of number concepts and operations and the creation and investigation of various computational algorithms. (For example, materials like multibase blocks can be used for whole number concepts; two-color chips and hot air balloons for integers; fraction bars for exploring rational numbers; and geoboards for certain irrational numbers.) In a similar way, area models, which play an important role in developing the meaning of multiplication and division, can be used to develop the definitions and properties of prime and composite numbers, ratios of numbers as probabilities, and irrational numbers.

Games, art, and market-place transactions provide an opportunity to help middle school students appreciate mathematics from a historical and cultural perspective. Prospective and practicing teachers, aware of various cultural and historical ties, can compare our numeration system with features and computational techniques of selected numeration systems from other eras and cultures. Such comparisons allow a distinction to be made between the concept of number and different ways of representing numbers and relate the uses of number to the culture in which it was or is used.

Computing technologies should be used to investigate methods of numerical computation and to explore and generalize number patterns and properties. Figurate numbers, Pythagorean triples, Fibonacci and Lucas numbers, Pascal's triangle, and sequences arising from geometric designs all provide suitable opportunities for prospective middle-grade teachers to explore patterns, make conjectures, and develop proofs.

Prospective and practicing teachers should investigate and develop an understanding of the relationship of fractions to repeating and terminating decimal representations, describe possible different algorithms for converting from one form to the other, and write convincing arguments to justify their conjectures. They should develop and present arguments for the irrationality of certain simple irrational numbers based on corresponding visual and decimal representations. They should grapple with a proof of the infinitude of primes and with other examples which demonstrate that different sets can have the same cardinality and also that sets of different cardinality exist.

Geometry

Geometry is a vibrant and exciting part of mathematics and a key to understanding our world. Recent research on the learning of geometry accentuates the need for concrete experiences with geometric figures and relationships prior to a formal axiomatic study of geometry. For middle-grade mathematics teachers, such concrete experiences are important not only in the development of their own geometric understanding but also in the enhancement of their knowledge of the stages through which geometric understanding evolves.

Standard 2: Geometry

The mathematical preparation of middle-grade mathematics teachers must include experiences in which they:

- ***investigate properties and relationships of shape, size, and symmetry in two- and three-dimensional space;***
- ***explore concepts of motion in two- and three-dimensional space through the investigation of rotations, reflections, and translations;***
- ***present written and oral arguments to justify conjectures and generalizations based on explorations;***
- ***become familiar with the historical development of Euclidean and non-Euclidean geometries.***

The geometry experiences of teachers of the middle grades should involve active participation, experimentation, and the use of different kinds of materials and models. These teachers need to build, explore, and extend a variety of geometrical patterns and describe them both numerically and algebraically. They can use wax paper folding to create conic sections or build polyhedra models and investigate their properties and relationships. They can analyze symmetries of designs and works of art from different cultures and learn to create their own designs. For example, they might create a template for drawing an Escher-type tessellation or use pattern blocks to build designs with certain symmetries. They might investigate properties of the golden rectangle and its relationship to the Fibonacci sequence, its use in art, and its appearance in nature. Geoboards can be used to give geometrical interpretations of irrational numbers and to develop alternative strategies, such as Pick's theorem, for determining the area of polygonal regions. Through the construction of models of buildings, bridges, and arches, they can explore questions about the strength, surface area, volume, or rigidity of the model.

Ideas of symmetry and patterns generated by movement provide opportunity for exploration. Experimentation in geometry is greatly enhanced by use of computer graphics or calculators with graphics capabilities. Prospective teachers might use Logo procedures to draw computer designs or create their own procedure to replicate a figure with a similar figure having specified properties. They might use the *Geometric Supposer* and software similar to it to explore properties and relationships of geometric figures. They can use graphing utilities to represent and study three-dimensional figures and analyze the properties of graphs.

Middle-grade mathematics teachers need to learn to reason about geometric figures using methods and results from transformation, coordinate, and synthetic geometry. Their experience in generating hypotheses (using exploratory computer software) should be followed with experience in establishing or disproving their conjectures. The focus, however, should be on the testing of hypotheses rather than on rigorous axiomatic development or proving theorems.

Exploring the geometry of the globe brings up questions about the meaning of the terms distance, line, parallel, perpendicular, similarity, and congruence. It forces a distinction between the properties of having finite length and being "endless." Finding triangles on the globe with three right angles forces a confrontation with our intuition about "what is true" about triangles. It sets the stage to discuss the development of non-Euclidean geometries. In the discussion of non-Euclidean geometries, there is no need for axiomatic developments or formal proofs. An informal, intuitive approach should be used to help middle school teachers develop a sense of history in one area of mathematics and an openness to the idea of creating mathematical systems that reflect different aspects of the world around us.

Algebra and Algebraic Structures

A basic understanding of functions and their applications is fundamental to the most important and common uses of algebra. Middle-grade mathematics teachers need experience both with the concept of function as used to describe real-world phenomena and with the basic elements of algebraic structures and their properties. In addition to enhancing their own understanding, this exposure to functions and algebraic structures will enable these teachers to help their students lay the groundwork that serves to build later connections to high school and college mathematics.

Standard 3: Algebra and Algebraic Structures

The mathematical preparation of middle-grade mathematics teachers must include experiences in which they:

- ***explore diverse examples of functions arising from a variety of problem situations and investigate the properties of these functions through appropriate technologies, including graphing utilities;***
- ***use physical models, charts, graphs, equations, and inequalities to describe real-world relationships;***
- ***explore and investigate properties of the integers, rational numbers, real and complex numbers (including order, denseness, and completeness);***
- ***use concrete examples to explore selected algebraic structures such as groups, rings, fields, and vector spaces.***

Prospective and practicing teachers of mathematics in the middle grades need many opportunities to build mathematical models that reflect realistic problem situations. Some problem situations will lead to mathematical models that can be recognized as particular algebraic systems in which properties of the system can be used to answer questions about the problem situation. Others will lead to various types of polynomial, trigonometric, exponential and logarithmic functions, Diophantine equations, difference equations, systems of equations or inequalities, and, possibly, matrix equations. Calculators and computers should be used to find exact or approximate solutions of equations, to represent equations and their solutions graphically, and to relate those solutions to the original problem situation. Strategies for finding solutions to problems should include guess and test, drawing pictures and diagrams, using manipulatives, making tables and charts, and translating relationships in situations to appropriate representations.

Modular arithmetic systems develop an appreciation of the idea of an algebraic structure. The arithmetic tables for Z_n, using several values for n, can be used to explore informally what it means to solve the equations $x + a = b$, $ax = b$, and $x^2 = b$; to determine the values of a and b in Z_n for which these equations have no solution, a unique solution, or multiple solutions; and to relate the number of solutions to the values of a, b, and n. Ring and field properties can, initially, be studied using modular arithmetic systems.

The question of existence of solutions can be used to motivate the extension of the natural number system to the integer, rational, real, and complex number systems. (The actual extensions should be done in a very intuitive and informal way.) The role of identities and inverses can be considered, and the properties of order, denseness, and completeness informally investigated. Questions of existence and uniqueness of solutions should be related to the problem situations which give rise to various functions and equations.

Examples of groups and subgroups can be built from investigations of the symmetries of triangles, quadrilaterals and regular polygons, and rotations of regular polyhedra or by looking for all possible arrangements of three different letters. Arrays of numbers used to record information provide an opportunity to introduce matrices and matrix operations.

Probability and Statistics

Statistical concepts, techniques, and results permeate our lives daily. The ability to make reasoned decisions in situations involving uncertainty is more important than ever. Concepts of probability and statistics are becoming increasingly prominent in the school mathematics curriculum at every level. Middle-grade mathematics teachers must provide their students with experiences that begin to develop the ability to reason about uncertain events. It is crucial that these prospective and practicing teachers develop solid, informal understanding of statistics from experiences involving the collection, organization, representation, and interpretation of data.

Standard 4: Probability and Statistics

The mathematical preparation of middle-grade mathematics teachers must include experiences in which they:

- ***collect data from experiments or surveys, organize and interpret data, and formulate convincing arguments based on appropriate data analyses;***
- ***make inferences and informed decisions based on statistical methods;***
- ***plan and conduct experiments and simulations to determine experimental probabilities;***
- ***develop counting and other techniques useful in determining theoretical probabilities;***
- ***identify the incorrect use of statistics by analyzing and critiquing arguments based on such incorrect use.***

In the study of both probability and statistics, emphasis should be placed on experimentation and data collection. Topics should be introduced by posing problems or situations to explore. Working in an experimental, small-group, activity-based environment, prospective and practicing teachers can discuss and share preliminary predictions about the problem or situation. Theoretical models and solutions should be presented only after the experimental approach has been fully explored.

Middle-grade teachers need to learn to read, construct, and interpret various tables, charts, and graphs that depict data. They should have numerous opportunities to use stem-and-leaf and box plots, plots over time, scatter plots, confidence intervals, survey analysis, and some informal hypothesis testing. They should learn to use the curve of best fit and to understand what it means for data to "fit" a particular curve. They need to recognize that not all data plot in a linear way. They should investigate connections between the binomial distribution and Pascal's triangle. Concepts of regression and correlation can be introduced informally. To integrate these ideas, teachers might do independent projects, investigating some topic or question of their choice in an area such as consumer product comparisons, environmental or health issues, sports, or opinion polls. The teachers should pose their own questions, gather and analyze relevant data, and prepare and present oral reports on their projects as well as a written report with statistical analyses.

These various experiences should include examples of misuses and appropriate uses of statistics and misconceptions about probability found in sources such as newspapers and magazines. Research has shown that people have well-entrenched, deeply rooted misconceptions about probability that are difficult to change. Through numerous probability experiments prospective and practicing middle grade teachers can come to recognize their own beliefs and build an understanding of what results are reasonable to expect.

Concepts of Calculus

The study of calculus makes a unique contribution to the mathematics background of any student. Through its study, students can develop an appreciation for the interrelatedness of powerful mathematical ideas. Middle-grade teachers must have opportunities to explore relations and concepts of change through models, concrete examples, and the use of calculators and computers. Calculus experiences for middle-grade mathematics teachers should arise from concrete examples and lead gradually to more abstract concepts.

Standard 5: Concepts of Calculus

The mathematical preparation of middle grades teachers must provide experiences in which they:

- *interpret, with the aid of graphs, diagrams, and physical models, the concepts of limit, differentiation and integration, and the relationship among them;*
- *construct concrete examples of finite sequences, extend the ideas to infinite sequences and series, relating them to the meaning of approximation of nonterminating decimals and to the approximation of functions;*
- *explore concrete realistic problems involving average and instantaneous rates of change, areas, volumes, and curve lengths, and relate these problems to the concepts of differentiation and integration.*

An introduction to calculus for middle-grade teachers should emphasize the underlying concepts of calculus, not technical skills. The teachers should actually construct concrete models to illustrate related changes (commonly used examples include making open boxes by cutting square corners from a given rectangular grid and comparing the change in volume to the change in height). They should use graphing calculators and computers to explore increasing and decreasing functions, to approximate local extrema and asymptotes of functions, and to investigate the behavior of discontinuous functions. Before introducing Riemann sums in a formal way, middle grade teachers should spend time estimating areas of irregular figures by counting squares of increasingly fine transparent grids. Irrational numbers such as $\sqrt{2}$ or π should be studied in terms of geometrical representations, in terms of limits of sequences, or as corresponding nonterminating decimal representations.

Prospective and practicing teachers can enhance their understanding of the computation of volume through calculus by first estimating the volume and surface area of a variety of figures. Similarly, string and wire can be used to develop an understanding of the definition of curve length. Middle-grade mathematics teachers should use calculators and computers regularly to graph functions, to calculate messy numerical quantities, and to generate patterns.

Teachers must have the opportunity, both in and out of class, to explore and develop the fundamental ideas of calculus through a variety of hands-on activities, small group discussions, and cooperative efforts on open-ended problems and questions. The basic properties of limits, derivatives, and integrals need to be developed in such a way that the students themselves can give good heuristic arguments for their reasonableness. For example, cardboard models, as illustrated below, can be used to represent the product of two functions and subsequently used to develop and justify the formula for the derivative of the product of functions.

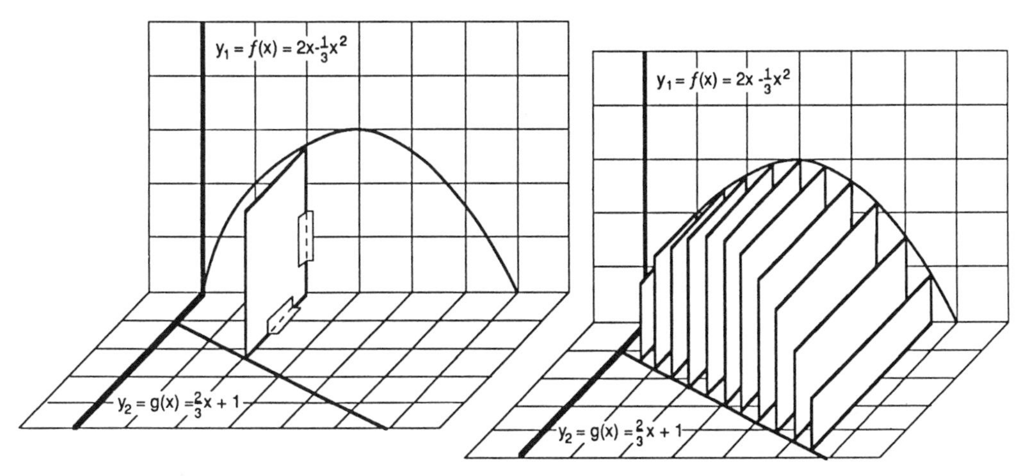

The value of the product of f and g at x is the area of the rectangle at x.

Section IV

Standards for the Secondary Level (9—12)

These recommendations assume that those preparing to teach mathematics at the 9—12 level will complete the equivalent of a major in mathematics, but one quite different from that currently in place at most institutions. The course work at the collegiate level is expected to meet the Standards in this section and those of Section I. The recommended Standards are:

1. Number Concepts and Properties
2. Geometry
3. Functions
4. Probability, Statistics, and Data Analysis
5. Continuous Change
6. Discrete Processes
7. Mathematical Structures

These Standards are not intended to describe specific courses, but rather to describe content areas in general terms and to capture the perspective on this content that we want students to understand. The specific topics covered are not as important as **how** those topics are taught. The Standards should be read as being independent of courses in the curriculum. Further, collegiate instructors of mathematics should anticipate, and might in the near future expect, that students in their classrooms will have had mathematical experiences during their secondary schooling which fully utilize the technological capability of calculators and computers. Consequently, we urge a deliberate, careful evaluation of how courses in the collegiate mathematics curriculum are taught, how they relate to one another, and especially to the Standards in Section I.

We recommend that all programs for prospective mathematics teachers at the 9—12 level include a year-long upper division sequence so that some area of mathematics is studied in depth. While there are many choices for such a sequence, the courses selected should require learners to explore some mathematical ideas deeply, make conjectures, justify them, and discuss the nature of their proofs. Course experiences should be designed to foster continued intellectual curiosity and to develop independence as a self-directed learner. Studies in analysis beyond introductory calculus are essential for prospective and practicing teachers who will teach Advanced Placement Calculus courses.

Mathematics teachers at this level should explore other disciplines in which mathematics is used. Course work from areas in the physical, biological, social, and behavioral sciences and in business and finance should be pursued so that teachers will encounter substantial applications of mathematics and will understand how problems arising in these disciplines have encouraged the creation of new mathematics.

Number Concepts and Properties

Secondary teachers need to explore number concepts and properties from a more advanced perspective than is provided by their previous high school or college experience with number systems (as described in Standard 1 of Sections II and III). The numerical techniques that have always been essential in using measurement and estimation to solve problems in all areas of applied mathematics have become even more important with the increasing use of technology. The fundamental limit concepts of calculus depend on thorough understanding of irrationality and the concept of infinity. Patterns of growth and rate of change are described through functions based on number relationships involving exponents and logarithms. The inductive nature of our number system is the stepping stone to inductive reasoning as a way of knowing and as a specific proof technique.

Standard 1: Number Concepts and Properties

The mathematical preparation of teachers of grades 9—12 must include experiences in which they:

- ***explore and discuss the properties, relations, and extensions of the natural numbers, integers, rational, real, and complex numbers;***
- ***employ an understanding of number concepts and properties to investigate mathematical concepts and applications in diverse settings;***
- ***study the historical development and significance of some major number-theoretic ideas and their applications.***

In extending some of the major number-theoretic ideas they developed earlier, prospective and practicing secondary teachers might study the Peano Postulates, investigate various properties arising in Pascal's triangle or the Fibonacci sequence, and explore some of the known results on Fermat and Mersenne numbers and on Euler's phi-function. Through suitably stated problems or group projects, the prospective and practicing teacher can become acquainted with the power of the Fundamental Theorem of Arithmetic and the division and Euclidean algorithms as they are applied in arithmetic investigations. Number theory problems are well suited to illustrate the interplay between calculator and computer use and the analytic techniques that can occur in learning and doing mathematics.

Investigations of the properties of various modular arithmetic systems or systems of matrices provide other settings in which numerical properties can be explored through both paper-and-pencil computation and computer-driven computation. In these investigations, emphasis should be placed on comparing and contrasting properties that hold in the familiar number systems with those in other systems. Applications to contemporary research areas like coding theory should be noted.

A course in elementary number theory or in the foundations of mathematics might suitably include activities described in this standard. An appropriate course in abstract algebra along with a real analysis course would also accomplish many of these objectives.

Geometry

Geometric ideas are playing a substantial role in today's world. Imaging techniques, string theory, minimal surfaces, knot theory, and fractals are just a few of the exciting new areas of study with geometric concepts as their basis. Because of the increasing variety of geometric applications, mathematics teachers need a thorough understanding of geometry, from synthetic, transformational, and algebraic perspectives and not limited to the plane, but including higher dimensions. This depth of understanding builds on the experiences described in Standard 2 on geometry included in Sections II and III.

Standard 2: Geometry

The mathematical preparation of teachers of grades 9—12 must include experiences in which they:

- *model features of the world in which we live using different geometries;*
- *develop geometric concepts both synthetically and algebraically using coordinates and vectors;*
- *use geometry as a source of mathematical models for a variety of applications;*
- *employ geometric reasoning as a problem solving strategy;*
- *become familiar with the historical development of non-Euclidean geometries and the questions relating to the parallel postulate involved in this development.*

Teachers need to develop conceptual understanding of the role of congruence and similarity in the classification of geometric figures. Special projects which involve a combination of exploration, building geometric models, research, proof, and written and oral presentations can bring together various conceptual ideas. The oral and written reports will help students organize their thinking and teach them to communicate effectively the main ideas of the project, its interest to others studying mathematics, its relationship to the course topics, and its applications in the "real world." Project experiences can be used to explore principles and examples of Euclidean and non-Euclidean geometries, transformation geometry, and applications of geometry to science and engineering as well as the arts. For example, projects developed around classical Euclidean construction problems can provide challenge for a broad spectrum of learners. Prospective and practicing teachers can draw upon such experiences in constructing their own further understanding of axiomatic systems.

The development of non-Euclidean geometries provides an opportunity for mathematics teachers to relate geometry to the surrounding world and to the history of mathematics. It sets the stage for asking questions about the parallel postulate, for looking at Saccheri's attempts to resolve these questions, and for describing the development of non-Euclidean and finite geometries in the last two centuries. These activities help prospective and practicing teachers develop a sense of history in one area of mathematics and an openness to the idea of creating mathematical systems that reflect different aspects of the world around us.

Teachers, given freedom to select some of the geometric investigations they pursue, develop examples of how to involve their own students in activities that are self-directed rather than teacher-directed. As an example, when studying geometric transformations, families of transformations generated by each of the following could be explored: (a) a single rotation; (b) a rotation and a reflection in a line through the center of the rotation; (c) a single translation; (d) two non-parallel translations. Through independent explorations, prospective and practicing teachers can discover the group structure for each of these sets of transformations and see some of the interrelatedness of algebra and geometry. Geometric designs incorporating the symmetry properties of these groups are helpful in visualizing the ideas. (Some examples might be drawn from corporation logos, pottery, wrought iron designs, woven rugs and blankets, wallpaper patterns, and Ukrainian Easter eggs. The art of Mauritus Escher provides an additional effective resource.) Prospective and practicing teachers should also work with current computer software designed for the exploration of geometric concepts.

A geometry experience, with attention paid to contemporary ideas of shape and dimension, is essential in the mathematical preparation of teachers. A course in college geometry might include many of the activities outlined here, as could a course in the history of mathematics. However, a combination of appropriately focused courses in linear algebra, graph theory, topology, differential geometry, and fractals could also include many, if not all, of the activities of this standard. Exploratory activities using software developed for geometric investigations could deepen students' understanding of conjecture in the theorem/proof process.

Functions

The concept of function is a central idea in mathematics and the sciences and arises in every course in the undergraduate curriculum. Mathematics teachers need to develop a solid understanding of this concept and its various applications. They need to understand the interrelationships among various methods of representing functions including graphs, tables, mathematical expressions, and verbal descriptions. Mathematics teachers also need to be aware that mappings, operators, affine transformations, and linear transformations are all functions—whatever the language, the underlying concept is the same.

Standard 3: Functions

The mathematical preparation of teachers of grades 9—12 must include experiences in which they:

- ***use the concept of function in the study of mathematics and the sciences;***
- ***represent functions as symbolic expressions, verbal descriptions, tables, and graphs and move from one representation to another;***
- ***use the language of functions to describe and model change;***
- ***investigate and discuss a variety of uses of functions in mathematics, business, and the physical, biological, behavioral, and social sciences.***

In the course of their collegiate experience, prospective and practicing teachers will encounter a tremendous variety of functions. They should be able to organize and classify these functions into broad groups such as continuous, discrete, periodic, differentiable, integrable, one or several variables. (For example, a logical circuit diagram can be represented as a table of input and output values. Such tables can be viewed as functions of several variables whose properties can be used to define, analyze, and correct a logical circuit. It is possible for students to develop a sequence of composite functions from these tables that could be used by design engineers to develop ways of connecting simple components to create complex logical circuits.)

Students should recognize that functions are mathematical objects and that, with suitably defined operations, functions of a given type form mathematical structures appropriate for further investigation. Prospective and practicing teachers can investigate algorithms and recursion as important illustrations of functions that have broad applications and interesting properties, especially to practical work in computing. From such investigations, students can begin to know and understand some of the ways that mathematicians use simple recursive functions to study chaotic behavior and fractals. The use of functions or operators to compare various mathematical structures should also be part of the mathematical experiences of prospective and practicing teachers. They should explore the role of "operation preserving" functions in identifying properties of related structures and in the building of new structures.

The fundamental nature of functions, the multiple representations that functions can have, and the variety of their different uses require thoughtful attention throughout the college mathematics curriculum. Courses in mathematical modeling and in applied disciplines give a breadth of knowledge of practical uses of functions that will provide prospective and practicing teachers with a sound background for teaching the function concept.

Probability, Statistics, and Data Analysis

Probability and inferences based on the knowledge of the probability of an event are important in statistical science. Prospective and practicing teachers of mathematics need to know various strategies to make reasoned decisions in situations involving uncertainty. Computer simulations provide a powerful tool to aid learners in understanding probability and in estimating solutions to problems with an element of chance. Prospective and practicing teachers need to develop an understanding of the distinction between experimental and theoretical probabilities through a variety of situations where they explore, discover, and work with this distinction in their college mathematics courses.

Standard 4: Probability, Statistics, and Data Analysis

The mathematical preparation of teachers of grades 9—12 must provide experiences in which they:

- *collect, display, analyze, and interpret sample data in a variety of situations;*
- *use experimental and theoretical probabilities as appropriate to formulate and solve problems involving uncertainty;*
- *explore the probabilistic nature of statistical analyses including hypothesis testing, correlation, analysis of variance, and some nonparametric methods;*
- *investigate the role of estimation and probability in statistical analysis, including various methods for estimating parameters and errors;*
- *develop strategies for reasoning and making decisions based on uncertainty.*

Mathematics teachers' knowledge must include familiarity with such basic probability distributions as the binomial, normal, Poisson, and chi-square. They will also need to develop and understand the concept of a random variable and its use in generating and interpreting basic probability distributions. Mathematics teachers should have adequate experience with probability theory to be able to deal with the relationship between correlation and cause, the difference between importance and significance, and the meaning of confidence intervals.

Decision making in the face of uncertainty is a central activity of modern life. Prospective and practicing mathematics teachers should understand the statistical and probabilistic underpinnings of the analysis of sample data so that they can, after collecting data in their own sampling experiments, better analyze, interpret, and make decisions based on those data. Computer software packages can be used to display information in tables, bar charts, pie charts, and box and whisker plots. The interpretation of such descriptive graphical representations of data can lead to reasonable decision making based on the data.

Statistical analysis of data can produce several types of information concerning the data which can then be used to make informed decisions about a situation. Experience with sampling from a variety of different databases using different sampling strategies and several statistical approaches will provide prospective and practicing mathematics teachers with the ability and confidence to present statistical concepts to their students at appropriate levels and in different contexts. Such experiences should also provide mathematics teachers with the ability to see misuses of statistics and to help students learn about such misuses and how to guard against them.

One approach to meeting the recommendations in this Standard might be a course in mathematical probability along with appropriate courses in data analysis, operations research, and mathematical modeling. To be able to deal effectively with large data sets in sophisticated ways it is assumed that courses used to satisfy this Standard have a strong computer component. If the intellectual and financial resources are not yet available in the department to provide each student access to, and instruction in, the use of substantial statistical computer software, the demonstration of statistical software packages in class and the use of statistical calculators in a mathematics laboratory are reasonable interim options.

Continuous Change

The measurement, description, analysis, and modeling of change is a fundamental activity in engineering, business, the sciences, architecture, economics, and mathematics itself. For example, the concept of derivative as a limit process, as a function, and as a rate of change permeates the investigation of continuous phenomena. It is through broad understanding and experience that prospective and practicing teachers can formulate and interpret various models of continuous processes. Teachers need to understand the relationship between differentiation and integration and the role continuity plays in this relationship.

Standard 5: Continuous Change

The mathematical preparation of teachers of grades 9 —12 must provide experiences in which they:

- ***use properties and techniques of calculus to model phenomena in diverse settings;***
- ***investigate the phenomenon of change as a limiting process;***
- ***explore both intuitively and in depth the concepts of limit, continuity, differentiation, integration and other continuous processes;***
- ***become familiar with the use of calculators with graphics capability and computer algebra systems both in the study and the applications of calculus.***

Any program preparing secondary teachers of mathematics will include a thorough introduction to calculus, and it is essential that current computing technology be included in that experience. In addition to learning the procedures, techniques, and standard applications of calculus, prospective and practicing secondary teachers must have an opportunity to come to grips with the fundamental underlying concepts of calculus.

Historically, while investigating continuous processes, many of the ideas and techniques of calculus were developed and used on an intuitive basis before the theory was made rigorous. Secondary teachers should have the opportunity to explore the basic ideas of calculus as described in Standard 5 of Section III. By building an intuitive base for analyzing continuous processes, these teachers might be more willing to take intellectual risks in their own classrooms. The actual material covered is less important than developing conceptual understanding of the ideas. The more detailed study of analysis will occur later in the major program.

In their collegiate experiences, prospective and practicing teachers should have the opportunity to model continuous real world phenomena from a variety of disciplines. This experience must include opportunities for them to formulate problems in mathematical terms and to interpret the meaning of solutions. Since some problems formulated in terms of calculus may be better approached using finite approximation techniques, both continuous and discrete techniques should be explored and compared. Problem situations which can be represented with sequences and series, for example, may be investigated through spreadsheets or recursive functions, but questions of convergence which use techniques studied in calculus may also arise. Applications of calculus may appear in a student's program in such courses as mathematical modeling, differential equations, applied or engineering mathematics, or numerical calculus.

Discrete Processes

In recent years there has been increased attention and use of discrete methods in mathematics and other disciplines. Problems from such diverse fields as business, industry, government, the natural and social sciences require finite processes and discrete techniques. The development of computer technologies has made possible exploration of practical problems which were otherwise too large or complex for analysis. Although the area called discrete mathematics is not yet well defined, the areas of mathematics commonly grouped under this term are an essential part of the mathematical preparation of secondary teachers.

Standard 6: Discrete Processes

The mathematical preparation of teachers of grades 9—12 must provide experiences in which they:

- *investigate a variety of problem situations which lead to diverse discrete mathematical models;*
- *develop and use a variety of counting techniques and counting arguments and their applications;*
- *gain experience in algorithmic and recursive thinking, and develop skills in using algorithms and iterative and recursive techniques in solving problems;*
- *learn to use appropriate technology effectively, including dealing with questions of computational efficiency and complexity;*
- *explore problems which involve the processes of scientific and social decision making.*

Prospective and practicing secondary teachers need to learn topics using discrete processes which are based on practical, realistic problems, on the mathematical modeling of these problems, and on the techniques involved in looking for solutions to the problems. Problems should be drawn from diverse fields including business, industry, government, and the natural and social sciences. This variety of problems can be used to introduce ideas found in operations research, the theory of finite graphs, linear programming, difference equations, recurrence relations, generating functions, matrices, and Markov chains. Specific examples might treat issues such as fair division, voting and election processes, apportionment, resource allocation, scheduling, and quality control. Some of the problem situations should lead to paradoxes or impossibility theorems. Others might lead to mathematical models in which there is no unique solution so that criteria must be determined for making choices from among several equally valid options.

Prospective and practicing teachers should see the use of matrices as a tool to represent the information contained in a finite graph, to describe recursive geometric transformations, to maintain inventory and financial records, and to predict steady state situations in Markov processes.

Counting techniques constitute a valuable strategy in addressing a variety of problem situations. Mathematics teachers should be able to use permutation and combinatorial computations in problems arising from several areas, including geometry, algebra, and graph theory. They should also understand how counting techniques apply in the calculation of the probability of events.

Use of technology in the learning and doing of mathematics enables learners to appreciate the power of simulation as a means to estimate the solution to problems involving uncertainty. It can also enhance their exploration and understanding of the role of iteration and recursion in the development of counting strategies and counting formulas. Analyzing and comparing specific algorithms, addressing questions of computational complexity and efficiency, and discussing the limitations of available technology are crucial to the effective use of these technologies in the search for solutions to problems.

To meet the experiences recommended in this Standard, some departments offer a discrete mathematics course or sequence at the elementary or more advanced levels. Others offer such courses as combinatorics, graph theory, difference equations, probability, or stochastic processes. Alternative approaches may be to require discrete structures courses in computer science or to include discrete topics in courses such as mathematical modeling or numerical techniques.

Mathematical Structures

The study of mathematical structures provides opportunities to see the beauty, elegance, usefulness, and efficiency of abstract mathematics. It allows prospective and practicing teachers to discover the fundamental essence of mathematical modeling, that of selecting and focusing on certain aspects of a problem situation; of relating those aspects to something already known in mathematics, often couched in terms of some abstract mathematical structure; and of using what is known about that mathematics to solve the problem. This approach must pervade all study and use of mathematical structures.

Standard 7: Mathematical Structures

The mathematical preparation of teachers of grades 9—12 must provide experiences in which they:

- *use mathematical structures which arise in the mathematical modeling of problem situations to explore and solve the problems;*
- *investigate mathematical structures that unify the observed patterns and properties that are shared by several diverse concrete examples;*
- *relate properties derived logically from the defining characteristics of a mathematical structure to properties of specific examples of the structure;*
- *explore the processes involved in building new structures from given structures (e.g., substructures, product structures, and quotient structures) and investigate properties and uses of such structures.*

The diverse areas in which given mathematical structures arise should be investigated and explored throughout the mathematical preparation of prospective and practicing secondary teachers. In areas as diverse as building error-correcting codes, studying DNA, solving scheduling problems, and analyzing artistic designs, a knowledge of mathematical structures and their properties is critical. Prospective and practicing teachers should develop an understanding of such structures as groups, polynomial rings, finite and infinite fields, vector spaces, topologies, and function spaces through a wide array of different problems and examples. Because of its wide range of applications, linear algebra should receive special attention in the preparation of secondary teachers.

Through their own investigation of these examples, prospective and practicing teachers can observe the shared properties which lead to the definition of such structures. In this way, they can directly develop an understanding of the interrelatedness of mathematical ideas. Deriving the properties of a mathematical structure from a definition, and relating the properties to those of other structures and to applications, provides opportunities for learners to reason mathematically, to make and test conjectures, and to support conclusions by oral or written argument. Similarly, in understanding the process involved in building new structures from given structures, prospective and practicing teachers can explore relationships among structures as they develop independent thinking and the confidence for doing mathematics on their own.

The introduction of definitions and theorems should be problem based and example generated. Prospective and practicing teachers need to see why the structures are of interest and why the definitions and theorems make sense so they can integrate these ideas with their prior knowledge. Before being introduced to the definition of isomorphisms, for example, prospective and practicing teachers should investigate numerous specific examples of groups and explore what might reasonably be meant by the groups being "essentially the same." With time to explore and discuss this idea they may formulate their own appropriate definition. Use of physical materials appropriate for the exploration of operations on integers and polynomials can be extended to explore properties of $Z(\sqrt{2})$ or the Gaussian integers.

Courses in abstract and linear algebra are important for prospective and practicing secondary teachers to study specific mathematical structures. But just as important are courses in mathematical modeling, differential equations, topology, number theory, analysis, and history of mathematics in which the role played by mathematical structures can be clearly emphasized.

Section V

Closing Remarks

Mathematics has long provided the language and computational techniques for science and technology. Today, people with mathematical ability and sound mathematical preparation are sought by employers in almost all fields. New developments in the physical, biological, social, and behavioral sciences, as well as business and industry, make preparation in mathematics essential for an adequately prepared work force. At the same time, the ways of "knowing and doing" mathematics are changing as well. Future citizens will be asked to approach and solve problems in ways quite different from those learned in their formal schooling. Hence, the mathematical preparation for students of tomorrow requires emphasis on understanding the conceptual bases of mathematics, an ability to communicate mathematical ideas to others, the ability to reason mathematically, and familiarity with the use of various technological tools in learning and doing mathematics.

In this document we emphasize that, although skills in mathematics must be acquired, processes, concepts, and understanding should take precedence. To accomplish this, the teaching of collegiate mathematics must change to enable learners to grapple with the development of their own mathematical knowledge. As we rethink the collegiate curriculum in mathematics, we must be open to new ways of presenting mathematical ideas. The standard curriculum in place for the past several decades should give way to a curriculum that weaves mathematical strands together to create new courses and new approaches to the development of ideas. To reenforce for learners that mathematics is a dynamic human endeavor, we need to incorporate some present day developments in mathematics. Technology is, and will be, a vital force in mathematics education. Its advent has changed the way we think about and do mathematics. Technology must influence our thinking about curriculum, for it changes the way we learn and teach mathematics in significant ways.

The use of computing technology in the teaching of collegiate mathematics is an exciting and necessary endeavor. Its full implementation requires time, energy, and commitment on the part of college teachers, but must be supported by a similar commitment on the part of college administrators. The value of technological change in the classroom must be recognized through promotion and tenure, space allocation, equipment and software procurement, and maintenance.

Ongoing professional development is critical for all teachers of mathematics, particularly in this time of rapid change in mathematics education. College and university faculty, working cooperatively with practicing teachers, can develop, both in the schools and in colleges and universities, a heightened awareness of issues that center around what mathematics we should be teaching, how we come to know and understand mathematical ideas, and how we can best teach to foster this kind of learning. In their efforts to take on this role, college and university faculty may seek advice and guidance. We urge the Mathematical Association of America, through its various committees, to meet this request by providing suitable support through appropriate publications, workshops, minicourses at national meetings. We urge the MAA Sections to undertake similar activities. We also urge dialogue with those responsible for graduate education for future mathematics faculty at the college and university level.

As teachers of collegiate mathematics, each of us has always thought deeply about what we teach. This call for change requires us to think equally deeply about **how** we teach. Because of this, we pose the following series of questions to consider each time we prepare to teach a course or plan a class. We feel that these questions are vitally important in enhancing instruction for all students.

1. What are the best ways to engage students in constructing their own knowledge, in expressing mathematical ideas orally and in writing, and in helping students grow in these abilities?

2. What are the most important mathematical concepts, processes, and results for students to learn? How will I communicate the importance of these ideas?

3. Are there better ways of introducing essential subject matter than by the traditional lecture method?

4. How can I convey my enthusiasm and excitement about knowing, learning, and using mathematics?

5. How can students be most actively involved in learning in the classroom? How can I demonstrate how to think through a problem to reach a solution? How can I convey the fact that multiple approaches to the solution of a problem are often possible?

6. How can this subject matter be developed from different perspectives and in several alternative ways so as to accommodate students with different backgrounds and learning styles?

7. How can this subject matter be connected to students' prior learning? How can this subject matter be related to the school mathematics curriculum and to other topics, especially those in other disciplines?

8. How can the subject matter be discussed in a way that emphasizes its importance both historically and in present day research and application?

9. How can I assess what the students already know or don't know? How can the development of the subject matter be monitored to assess whether students are gaining the expected understandings and skills?

10. Aside from traditional testing, what other methods can be employed to assess what students know and are able to do?

11. How can the contributions to the subject matter of this course that have been made (or are being made) by women or minority mathematicians be highlighted?

12. In what ways can calculators or computers be appropriately used to enhance students' understanding of the subject matter? In what ways do calculators or computers alter the selection of ideas that should be stressed in this course? How do they change the ways of assessing students' understanding?

13. Has a broad choice of relevant examples been employed to motivate and illustrate the mathematical concepts presented? Do the examples illustrate diversity in gender, ethnicity, and culture?

14. Will students take away from the course not only specific knowledge and skills in the subject matter but also an appreciation for the growth and development of mathematics as a discipline?

15. What techniques can be employed to foster the joy of learning mathematics and exploring conceptual ideas? How can a positive attitude toward mathematics be developed in the students?

16. What topics are particularly appropriate so students can experience the beauty and fascination of mathematics?

Already, in many classrooms across the country, the spirit of the NCTM ***Standards*** is taking hold. It is the urgent responsibility of college and university departments of mathematics to provide prospective and practicing teachers with experiences that enable them to acquire mathematical knowledge, to understand the concepts and processes inherent in mathematics, to see the interrelations among the various branches of mathematics, to recognize the relationships of mathematics to other disciplines, to feel confident in their ability to do mathematics, and to have an appreciation of the power, beauty, and fascination of mathematics. This is the call for change, a vital, important, and necessary change that will benefit us all.

References

A Source Book for College Mathematics Teaching, 1990, MAA Committee on the Teaching of Undergraduate Mathematics, The Mathematical Association of America, Washington, DC

Calculus for a New Century, MAA Notes Number 8, 1988, The Mathematical Association of America, Washington, DC

Challenges for College Mathematics: An Agenda for the Next Decade, Report of a Joint Task Force of the Mathematical Association of America and the Association of American Colleges, MAA *Focus*, November - December, 1990

Curriculum and Evaluation Standards for School Mathematics, 1989, National Council of Teachers of Mathematics, Reston, VA

Everybody Counts: A Report to the Nation on the Future of Mathematics Education, 1989, National Academy Press, Washington, DC

Mathematics Education: Wellspring of U.S. Industrial Strength,
1988, National Academy Press, Washington, DC

Moving Beyond Myths: Revitalizing Undergraduate Mathematics, 1991, National Academy Press, Washington, DC

On the Shoulders of Giants: New Approaches to Numeracy, 1990, National Academy Press, Washington, DC

Priming the Calculus Pump: Innovations and Resources, MAA Notes, Number 17, 1990, The Mathematical Association of America, Washington, DC

Professional Standards for Teaching Mathematics, 1991 National Council of Teachers of Mathematics, Reston, VA

Renewing U. S. Mathematics: A Plan for the 1990s, 1990, National Academy Press, Washington, DC

Reshaping College Mathematics, MAA Notes Number 13, 1989, The Mathematical Association of America, Washington, DC

Reshaping School Mathematics: A Philosophy and Framework for Curriculum, 1990, National Academy Press, Washington, DC

Toward a Lean and Lively Calculus, MAA Notes Number 6, 1986, The Mathematical Association of America, Washington, DC

Appendix A

College and University Responsibilities for Mathematics Teacher Education

(Approved: January, 1986, MAA Board of Governors)

College faculty must become actively involved in the education of teachers if the teaching of mathematics in the schools is to improve significantly. Active leadership and support of college and university mathematicians, mathematics educators, and administrators is essential if our nation is to increase the number of qualified teachers and to strengthen their education. For this reason, the Mathematical Association of America and the National Council of Teachers of Mathematics have adopted the following recommendations for all individuals, in whatever department, who are engaged in teaching mathematics or mathematics education for current or prospective teachers:

1. Colleges and universities should assign significantly higher priority to mathematics teacher education.

2. All individuals who teach pre-service or in-service courses for mathematics teachers should have substantial backgrounds in mathematics and mathematics education appropriate to their assignments.

3. Mathematics methods courses should be taught by individuals with interest and expertise in school teaching and continuing contacts with school classrooms.

4. All individuals who teach current or prospective mathematics teachers should have regular and lively contact with faculty in both mathematics and education departments, e.g., by regular meetings, seminars, joint faculty appointments, and other cooperative ventures.

5. All college and university faculty members who teach mathematics or mathematics education should maintain a vigorous dialogue with their colleagues in schools, seeking ways to collaborate in improving school mathematics programs and in supporting the professional development of mathematics teachers.

6. Faculty advisors should encourage their mathematically talented students to consider teaching careers.

7. Colleges and universities should vigorously publicize the need for qualified mathematics teachers and strive to interest and recruit capable students into the profession, e.g., by organizing

highly visible campus-wide meetings for students to inform them of the opportunities, advantages, disadvantages, and requirements of a career in teaching mathematics.

8. Tenure, promotion and salary decisions for faculty members who teach current or prospective mathematics teachers should be based on teaching, service, and scholarly activity that includes research in mathematics or mathematics education.

9. Faculty members in mathematics and in mathematics education who are effective in working with activities in the schools and in the mathematical education of teachers should be rewarded appropriately for this work.

10. All institutions involved in educating mathematics teachers should provide specialized classroom and laboratory facilities equipped with state-of-the-art demonstration materials, calculators, and computers at least comparable to those used in the best elementary and secondary schools so that prospective teachers, like graduates from other professional programs, can be properly prepared for their careers.

Appendix B

Previously Published Recommendations for the Mathematical Preparation of Teachers by CUPM and COMET

The Committee on the Mathematical Education of Teachers (COMET) was established in 1983 by the Mathematical Association of America (MAA) to succeed the Panel on Teacher Training of the MAA's Committee on the Undergraduate Program in Mathematics (CUPM). From its beginning more than thirty-five years ago, CUPM has been concerned with the mathematical preparation of elementary and secondary school teachers. COMET continues this tradition both for CUPM and for the MAA.

Recommendations for the Training of Teachers of Mathematics (1961; revised 1966)

Course Guides for the Training of Teachers of Elementary School Mathematics (1961; revised 1968)

Course Guides for the Training of Teachers of Junior High School and High School Mathematics (1961)

Teacher Training Supplement to the Basic Library List (1965)

Recommendations on Course Content for the Training of Teachers of Mathematics (1971; Reprinted in A Compendium of CUPM Recommendations, Vol. 1, pp. 158-202)

Recommendations on the Mathematical Preparation of Teachers (1983; MAA Notes Number 2)

Guidelines for the Continuing Mathematical Education of Teachers (1988; MAA Notes Number 10)

Related Publications

A Call for Change addresses recommendations for the mathematical preparation of teachers. Related publications of the National Council of Teachers of Mathematics, the Mathematical Sciences Education Board, and the National Research Council are important for the context of this document and the goals of improving mathematics learning for all students.

From the National Council of Teachers of Mathematics

- *Professional Standards for Teaching Mathematics* (#439S3)
- *Curriculum and Evaluation Standards for School Mathematics* (#396S3)

Single copies of each of the above are available at $25 each. To order, or inquire about multiple copy purchase, contact:

The National Council of Teachers of Mathematics
1906 Association Drive
Reston, VA 22091

Phone: (703) 620-9840 or, for orders only, (800) 235-7566

From the Mathematical Sciences Education Board

- *Counting on You: Supporting Standards for Mathematics Teaching*

Single copies are available at $2.95 each. Orders of ten or more: $2 each.
To inquire about larger quantity discounts, contact MSEB.
Payment, by check payable to MSEB, must be enclosed with order.

Mathematical Sciences Education Board
TPROF
818 Connecticut Avenue, NW, Suite 500
Washington, DC 20006

From the National Research Council

- *Everybody Counts: A Report to the Nation on the Future of Mathematics Education*
- *Moving Beyond Myths: Revitalizing Undergraduate Mathematics*

For prices and other order information, contact:

National Academy Press
2101 Constitution Avenue, NW
Washington, DC 20418
Phone: (800) 624-6242 or (202) 334-3313

SMCL
3 5151 00163 1704

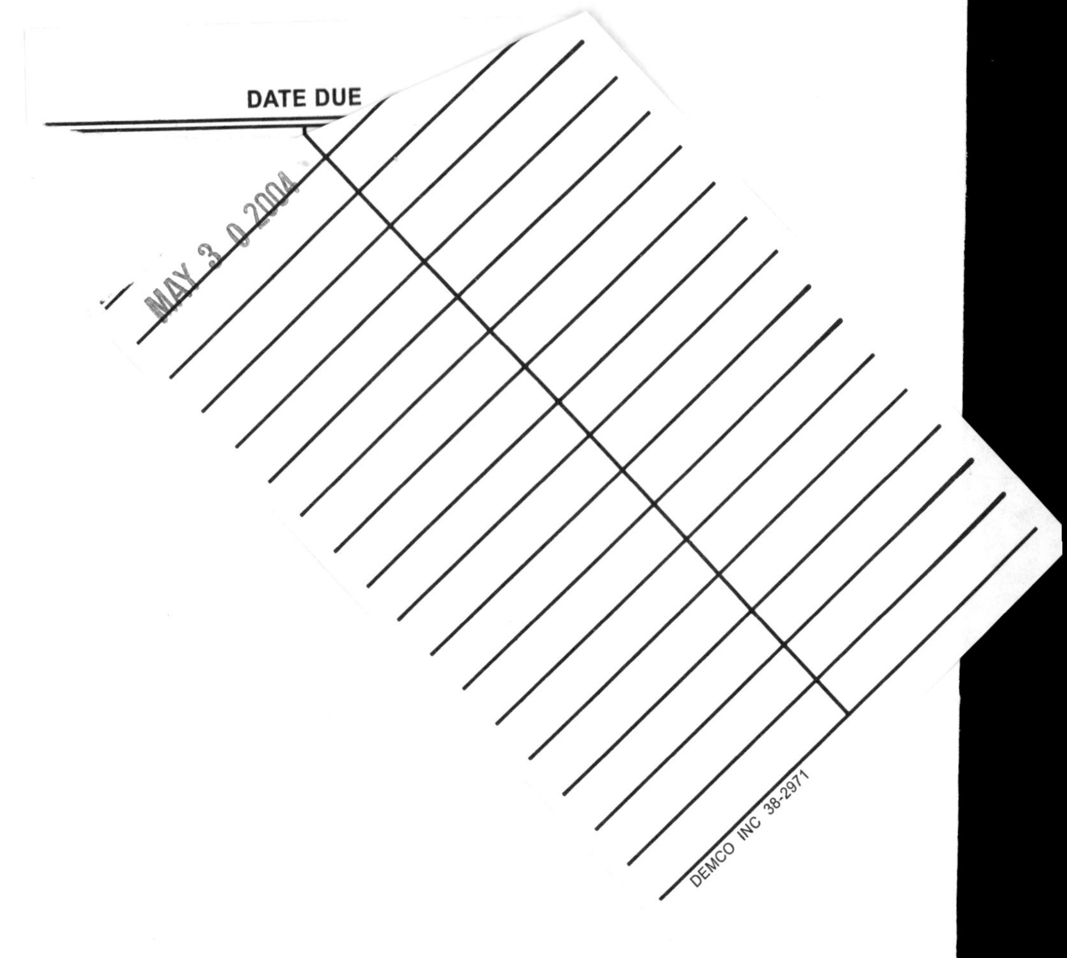
DATE DUE
MAY 3 0 2004
DEMCO INC 38-2971